# A Lifetime Burning

# A Lifetime Burning

### Richard Quinney

Copyright ©2010 Richard Quinney

All rights reserved. No part of this book may be reproduced in any form without written permission from the publisher, with the exception of brief excerpts for review purposes. Published by Borderland Books, Madison, WI www.borderlandbooks.net

Publisher's Cataloging-In-Publication Data

Quinney, Richard.

A lifetime burning / Richard Quinney. — 1st ed.

ISBN: 978-0-9815620-5-6

1. Quinney family. 2. Family farms — Wisconsin — History. 3. Farm life — Wisconsin — History. 4. Quinney, Richard — Childhood and youth. 5. Wisconsin — Biography. I. Title.

CT275.Q554 L55 2010

977.5/092                                       2009925843

Printed in the United States of America
First edition

Design: Ken Crocker
Printing: Capital Offset Company, Inc.
Binding: Acme Bookbinding

Cover and illustrations from the childhood scrapbook of Alice Quinney, given to her son Richard Quinney on his first birthday.

Home is where one starts from. As we grow older
The world becomes stranger, the pattern more complicated
Of dead and living. Not the intense moment
Isolated, with no before and after,
But a lifetime burning in every moment
And not the lifetime of one man only
But of old stones that cannot be deciphered.

– T. S. Eliot, "East Coker"

**The farmhouse is empty now.** The accumulated effects of four generations have been cleared out of the house. A century and a half of life on this midwestern farm.

Artifacts from the previous generations had been stored in the house as the generations passed. In the attic were the spinning wheel, the highchairs, the baby beds, the quilts, the cradle scythe, boxes of books, framed paintings and photographs, sets of dishes and silverware, and, strewn on the floor, the rosary beads that served my great-grandmother Bridget for a lifetime. The basement held the tools, the sleds, the milk cans, barn jackets, caps, the workbench, and the canning jars. On the front porch were the emigration trunks that held scrapbooks, photographs, farm ledgers, diaries, and the souvenirs from trips to the West. The cupboards, dressers, and closets were filled with the material things from the years of living by my mother and father, and from the early years of my brother and me. Every object uncovered held a story and a memory. Every night after a day of sorting and clearing at the farmhouse brought fitful dreams and sudden awakenings.

With great care I removed the small oil lamp that my father carried as he climbed up the stairs to bed each night when he was a boy. All these years the oil lamp had been on a shelf in the farmhouse. Taking the lamp from the house would be my farewell to the place I have always called home. Now the lamp rests on the high shelf of the corner cabinet in my house in town. The lamp still lights the way.

***B**efore taking the brown, tattered scrapbook* down to the basement of my house for storage, not yet willing to give up this earthly possession, I paged through it one more time. Inside the

## A Lifetime Burning

cover was the inscription in my mother's hand: "To Earl Richard Quinney, From Mother, 1935." This is the scrapbook that my mother made when she was a young girl and gave to me the year after I was born so that someday I might learn something about her childhood. From front to back the pages are filled with gift cards, valentines and birthday cards, pictures clipped from magazines, stamps, bird and flower stickers, and labels from tins and jars, all carefully pasted to the pages. Artifacts, now, of things that once my mother valued and saved and passed on to me.

**Some of the treasures** from the farmhouse will be preserved by the state and local historical societies. The old furniture and family heirlooms have been dispersed to the homes of children and grandchildren. The kitchen table that our family sat around is now in my dining room and serves as my writing table each morning. Some material objects of daily living are gone forever, such as my mother's green ceramic teapot. Still, my desktop is covered with enough knickknacks to last the rest of my life.

Truly, as T. S. Eliot wrote in his poem "East Coker," "Home is where one starts from." And the rest of the lines that I have chosen as the epigraph are true to my life as well. "As we grow older / The world becomes stranger, the pattern more complicated / Of dead and living. Not the intense moment / Isolated, with no before and after, / But a lifetime burning in every moment." Gracefully given, a lifetime burns every moment in these late years.

*Maybe it is in the genes,* certainly in the nurturing by generations of farmers in my family. Born—and raised—on the farm is central to who I am. My great-grandfather Quinney was an Irish peasant farmer in County Kilkenny. My great-grandfather Holloway, on my mother's side, was from a family of tenant

## A Lifetime Burning

farmers in the west of England. Each generation after the emigrations of the 1840s and '50s farmed the land.

My father did not encourage me to stay on the farm; in fact, he told me I would have a better life doing something else. Working in the fields, he would sing, "How 'ya gonna keep 'em down on the farm after they've seen Paree?" After you have traveled, as my father once did as a young man, you'll "never want to see a rake or plow, and who the deuce can parleyvous a cow," the song continues. In his midlife I heard my father say that he would like to move to town and open a restaurant.

I would not have stayed on the farm even without my father's advice. I wanted to be someone other than a farmer. Someone, I thought, of the "modern" world, sophisticated and worldly. Most of all, I wanted to be part of a world beyond the country life I had known as a child of the late 1930s and the 1940s. But here I am, more than half a century later, trying to preserve the farm. I am an inheritor of the farm, yes, a weekend

occupant and an advocate of the future of family farming. I may have left the farm, but the farm never left me.

***I might never have cleared the farmhouse*** of the material possessions of the past — those of my great-grandparents, my grandparents, my mother and father, and my generation. A museum it could have been if I had decided to make this my home in the country. But age and the desire to be at home in my house in town prompted me to clear the farmhouse and make room for others.

Renters surely will make their own lives in the farmhouse that I have abandoned. Of the family possessions, the upright piano — too heavy to move — is all that remains. Music will still fill the house.

The farm buildings, including the large barn, stand in various conditions of ruin and repair. The 160 acres of farmland, woods, and marsh continue to be maintained and preserved. Other generations will determine the course and fate of the land that has supported the previous generations of the family. Methods of sustainable agriculture are being practiced and demonstrated on these acres. The whole farm, as envisioned, is a natural habitat. I will

## A Lifetime Burning

watch, now, with only the miles between farm and town separating us. Here, on yet another border, I remain, a farmer. Fortunate and happy am I for this life.

**The weather always connects us** to our essential existence. The morning newspaper offers the forecast: Bitterly cold air, accompanied by strong winds from the north, will surge from the northern plains into the northern Mississippi Valley. Low pressure developing along the arctic cold front will cause an area of snow to expand from Iowa and eastern Minnesota this afternoon to the western Great Lakes states tonight. Several inches of wind-driven snow will accumulate, especially tonight over Wisconsin.

Reminding me again, with great pleasure, of the last lines from James Joyce's story "The Dead": "Snow was general all over Ireland." As Gabriel watched from the window of the hotel, Joyce writes, "his soul swooned slowly as he heard the snow falling faintly through the universe and faintly falling, like the descent of their last end, upon all the living and the dead." I too am with soul swooning slowly, sitting here at my table, snow falling faintly through the universe. The temperature will fall to twenty degrees below zero tonight. Everything is in season. Appearances aside, we are on schedule. The end of the weather report focuses on the larger perspective: Though masked in some

## Richard Quinney

areas by deep snow and biting winds, the slow climb from winter into spring has begun. Average temperatures will start to rise across the country. Daily weather may obscure what is actually happening. I will write daily, among other reasons, to obscure the coming of the inevitable fate. No life without death. Or maybe, beyond our illusions, the world is something other.

***I have had this need to write,*** at least in a notebook, all my adult life. An attempt to find out who I am, to put myself together anew each day. My need to make a connection with others near and far.

Clearing out the drawers and trunks at the farm, I uncovered writings by the generations before me. I found the childhood diaries of my mother, the letters of my father, the farm ledgers and diary of my grandfather, the notes of my great-aunt Kate, the autograph book of my aunt Marjorie, the notebook of my great-grandfather written on board the emigration ship to the New World. These are the personal writings that have survived,

## A Lifetime Burning

while others likely have been lost forever. The writings that survive connect us to one another, the living and the dead. No before and no after, as T. S. Eliot wrote, but lifetimes of many burning in every moment.

**Not so long ago,** only days after my mother passed away, I found in the desk drawer the notes she had made near the end of her life. She had written them for herself, certainly, but also for us, to tell us who she was and how she might be remembered. The only child of a farmer and his wife, she grew up in the country. The handwritten list is on a piece of note paper:

> *Grew up in an adult world.*
>
> *Sheltered and protected.*
>
> *Discipline fair and loving.*
>
> *Difficult birth and Dr. advised my parents not to have any more children.*
>
> *Center of my parents' world.*
>
> *Whenever my parents went to visit friends, a ride in the buggy or walks, fishing, trips to Geneva Lake or to town I went too.*
>
> *In school I sometimes felt left out.*

## Richard Quinney

In her handwriting, on the back of the note paper, she copied "Wesley's Rules":

*Do all the good you can*

*By all the means you can*

*In all the ways you can*

*In all the places you can*

*At all the times you can*

*To all the people you can*

*As long as you can.*

My mother added, at the end of the page, the meaning of her name, Alice: "Truthful One."

## A Lifetime Burning

**A *leather-bound diary of my grandfather*** Will Holloway, my mother's father, for the years 1906 and 1907 was at the bottom of the trunk. I turned immediately to the entry for April 29, 1906, the day my mother was born, the birthday that she would celebrate for the rest of her life. My grandfather had written: "Pleasant. Dr. Caffee, Miss Keopke, Miss Christianson were here. Baby was born at 11:00 a.m. Thos. Taylor and G. Rhodes were buried." The words "baby was born at 11:00 a.m." were underlined. The entries throughout are brief and concise, referring to the weather, field work, trips to town, cost of supplies, sale of products, visits with relatives, funerals. Never is there an emotion explicitly expressed or a contemplative thought rendered. Still, a diary full of life. Of the life lived — and observed — daily.

***In the desk drawer,*** with the notes my mother made, was a description of her father that she had written at the same time she described herself.

> My father was a man of many interests. He grew up on a farm. He rented his father's farm and owned 40 acres of his own in

## Richard Quinney

*Sugar Creek Township. He milked cows, raised sheep, chickens and registered Duroc Jersey pigs. He worked the land to provide food for the animals. He enjoyed gardening and always raised potatoes, onions and squash to sell. In 1920 he retired from farming and moved to Millard. He kept busy doing various jobs, land measuring, spray painting, going back to the farm doing whatever repair jobs needed to be done. He was appointed guardian for several people and was administrator for some estates. For 56 years he was the Town Clerk of Sugar Creek.*

No words were found in the drawer about my mother's mother, who had died when my mother was a young girl. But my mother always kept a framed photograph of her mother on the table in the living room.

**Early one morning** we drove to the farmhouse to take my mother to an appointment in town with her doctor. She had

## A Lifetime Burning

waited for us throughout the night in the living room chair, and as we were helping her prepare to leave the house she passed away. No amount of writing here, now after several years, will ease the loss, nor offer much solace. I wake up in the night often with great sadness thinking about, and missing, my mother and father. Writing does not lessen the grief of the loss. I write the words as a way of going on. An answer of sorts to the silence. A silence burning in every moment.

***A cautionary tale:*** Zen Buddhist monk Ryokan said that he would tell us a secret: "All things are impermanent." Within quotation marks because the words are common to us, the idea so familiar, but we live our lives to the contrary. Our thoughts would have it otherwise, that things are permanent. But our experience is that of things always coming into existence, changing, and disappearing. All objects of thought and perception have no independent and permanent existence of their own. Buddhists call this reality "emptiness."

How are we to live with the insight of emptiness? I repeat and remind myself as often as I can of the wise and concise statement from the introduction to the book *Zen: Images, Texts, and Teachings:* "The paradox of everything we experience in life is that it is truly empty and at the same time possessed of a marvelous subtle, mysterious existence. Everything is empty, yet spring comes, flowers bloom, and trees show new growth. Everything is empty, yet even the most ordinary thing is marvelous itself. Zen masters teach that to realize the emptiness and interconnectedness of all things, not just with the mind but with one's whole being, is to achieve enlightenment — to become a Buddha."

But even with this insight, we long for permanence, trying to preserve the past. Such seems to be our human condition. My desire to preserve the past, to hold onto the memory, prolongs the sadness of loss. I find comfort and solace when I focus on the present moment, on the wondrous things that are happening

## A Lifetime Burning

now. The words of the fourteenth-century Taoist poet Ch'ing Kung bring me to attention:

*Things of the past are already long gone*
*and things to be, distant beyond imagining.*
*The Tao is just this moment, these words:*
*plum blossoms fallen; gardenia just opening.*

Just paying attention to the reality of the interconnectedness of all things and the impermanence of this existence.

**M**usic comes through the speakers, filling the living room. A new song by Willie Nelson, "Always Now," furnishes an answer to the quandary of loss and living in the present. "It is always now/And nothing ever goes away/Everything is here to stay/And it's always now."

Take heart in the insight that nothing is ever lost, that lives and things of the past are with us always. Everything that ever existed is here in the present. The emigration trunk that was stored on the front porch all those years is still with me, even though it is elsewhere, at the local historical museum. My ice skates that hung for sixty years in the attic of the farmhouse have been discarded, but they still are a part of me as I live each day. Parents and ancestors and friends are still with us. All things of the past are here to stay. Gone forever is only a thought.

Impermanence, interdependence, insubstantiality are of the reality that there is no separation between past, present, and future. All is here in the present moment. Nothing is lost; everything is with us now. Yet, there is a sadness that will not go away.

**W**hat remains, materially, are the artifacts, the material things now devoid of their former purpose and function. We make decisions about their dispersal. The items that survive await some other use, maybe only as a reminder of the past and the

## A Lifetime Burning

transience of life and all things material. These old things may have an imaginary afterlife, the afterlife of things.

We who remain give new meaning to these things. We entertain the possibility that something—call it spirit—is contained in the material world of things. I sometimes sense a spirit moving in the rocking chair, now in my living room, that once resided beside the fireplace in the old house a hundred years ago. And who does not experience the presence of a spirit when looking into the eyes of an ancestor photographed long ago?

I know daily the wonder of material things. In this little room where I do much of my work, all surfaces—vertical and horizontal—are filled with things of various use and vintage. I remind myself of the photographer Josef Sudek in his studio in Prague sixty years ago. In an interview, Sudek said that he loves

# Richard Quinney

the life of objects. In his photographs he told stories about the life of inanimate things.

Sudek lived and worked surrounded by, immersed in, these material things. Sudek's last series of photographs were still life portraits of the material things found naturally, or arranged, in his studio. Photographs of things purchased cheaply at auctions of country estates and convents and monasteries, souvenirs, and objects given to him by friends. Piles of paper on tables, letters, catalogues and magazines, tools, balls of string, all composed in images alluding to a reality behind the object world.

In my own house, surrounded by the material objects of the past, I am transported to another world. Transported to a spirit world of sorts. After his death in 1976, Sudek's studio was dismantled with great care.

### A Lifetime Burning

**In our family,** we always thought that we were not far removed from the emigrations of our ancestors. The generations receded slowly from memory, but their leaving of the Old World for the new would stay with us, indelibly etched into our imaginations. Growing up on the farm, I heard the stories of the emigrations of our ancestors from England and Ireland. The stories certainly were shortened in the telling through the generations, yet sharpened and more impressive in the shaping of our minds.

For me, in the fourth generation, the image of being an emigrant and a pioneer came to define the course of my life. I might sometimes attempt to remove myself from the emigration stories, but the attempt only reinforced the history of those who came before me. To this day, daily, I too am from another place.

James Holloway, my mother's grandfather, booked passage for Australia and left the tenant farm in North Devon known as Bright's Leary. On the morning of departure, for reasons unknown to us, he changed his ticket and sailed to Quebec, for thirty days on the rough seas in the summer of 1865. Two years later, after James had migrated to Wisconsin, Mary Bray immigrated to Wisconsin from Devon with her father and mother and brothers

## Richard Quinney

and sisters. Mary and James would meet and marry and have two children, Will and Lizzie. They moved from farm to farm in Walworth County, finally purchasing the acres north of Millard. On a nearby farm, Will met Lorena Taylor, whose grandparents had emigrated from England in the 1850s, and she became his bride. Their one child, born Alice Marie in 1906, became my mother in 1934. I keep a framed photograph of my grandfather on my writing desk. Taken late in his life, it shows him at his desk, looking down at the plat map for Sugar Creek Township. Next to his photograph is a photograph of Lorena, in a long skirt, standing in the farmyard looking into the distance.

I was born on the farm that was settled by my great-grandparents John Quinney and Bridget O'Keefe, who had emigrated from Ireland during the potato famine of the 1840s. Meeting in Yonkers, New York, and soon marrying, they eventually made their way to Wisconsin and rented an acre of land south of Millard. In 1868, they purchased the few acres that would become the 160-acre farm that was my childhood home.

One of their five children, John, farmed the land all his life. In 1894 he married Hattie Reynolds, who had grown up on a farm near Milton in Rock County. They lived in the house at the old place and had three children, including my father, Floyd.

## A Lifetime Burning

Hattie was one of nine children of Nathan Church Reynolds and Mary Potter. Nathan's grandfather Daniel Reynolds had lived in Maine and Massachusetts and had fought in the Revolutionary War. His ancestors and those of his wife, Thankful, had also been immigrants to the New World.

After Hattie died of consumption in October of 1905, when my father was five years old, he was raised at the old place by his father and his grandmother Bridget. In 1929, at a dance in Delavan, my father met Alice Holloway. They were married the next year and built a new house on the farm. My brother and I heard the stories of our family—the emigrations, the migrations, the births, the deaths—as we grew up on the farm. The lives of our ancestors are still with us, now, in our daily lives.

**W**ater *drips from the ends of the long icicles* that hang from the eave of my house in town. As evening comes, the icicles will have grown even longer from the melting and the freezing. Warmer days will come, and the icicles will vanish from outside my window. The icicles will not have been in vain; the memory will last and give pleasure to the day in another season.

## Richard Quinney

    The truth of Ecclesiastes, again, is remembered. A lesson and a warning against attachment, pride, and arrogance. The revelation that this too will pass despite the strength of our human will and passion. "Vanity of vanities, says the Teacher, vanity of vanities! All is vanity." What is to be gained from our toil and from the wisdom we have gained? All things are impermanent. Generations come and generations go.

    The present is our time. Eternity is in the moment as I watch the water dripping from the ends of the icicles. An ultimate meaning of existence for the human mind to comprehend. A friend has just written to me that we are "spiritual beings having a physical experience rather than physical beings having a spiritual experience." Suggesting that the spiritual realm, and our place in it, survives. Any other hope would be another vanity among vanities. A vapor.

## A Lifetime Burning

**W***hy this continuing obsession* with the material possessions from the past? As we get older, the world does become stranger, and holding on to these possessions strengthens this strangeness. Certainly depth is added to our lives; but I know, at the same time, that there could be lightness to my daily living if I would shed some of these remains. My solution, for the time being, is to make these artifacts a part of my life. All of the past is contained in the present moment.

These material things—the artifacts of previous lives and times—are physically alive in the whirl of atoms within them. The electrons, protons, and neutrons within these materials are no different from those that make up the human brain. A rock consists of atoms connected by a pattern of chemical bonds. That rock is connected to the entire universe by the gravitational and electromagnetic signals continuously received. Analogous to what is happening in the human brain, albeit a consciousness of a different kind, but still a consciousness.

The term "panpsychism" is given to the ancient philosophy that mind exists, in some form, in all living and nonliving things. This idea, revived in contemporary philosophy, suggests that life is present and pervasive in all things. The artifacts that I now sort through, in fact, have a life of their own. A life beyond the part they play in my daily life.

**My daughter, home for a visit from afar,** and I sat at the dining room table and looked together at the tattered scrapbook that my mother and father made following their marriage in 1930. Pressed between the pages are the flowers and the ribbon with my mother's note: "My wedding bouquet—yellow roses—beautiful oh so beautiful." The clipping from the newspaper describes the wedding, announces the wedding trip to Niagara Falls and Canada and the plans to return "to the farm where a new house is awaiting them." The scrapbook contains the many cards of congratulation sent to Floyd and Alice by their friends and relations.

The scrapbook contains the anniversary cards and the souvenirs from vacation trips to northern Wisconsin. Pasted to the page is a brochure from "Paul Bunyan's Home Camp on Beautiful

## A Lifetime Burning

Rice Lake." We look at the birthday cards, the valentines, and the Christmas cards my mother and father once gave to each other. Terms of endearment, handwritten: "My darling husband," "Your hubby," "My dearest and only sweetheart." A love that lasted throughout their lives.

And then the announcement card appears in the form of a baby carriage. "Hello Everybody!" Inside the card, in my mother's hand, "I arrived May 16 to stay with Mr. and Mrs. Floyd Quinney. My name is Earl Richard." Two years later, the birth of my brother is announced. Cards of congratulation follow the births of the boys. Eventually, the boys are sending cards to their parents, cards that will carefully be saved. This was a family, ever so dear.

Upon returning to her home, my daughter responded sympathetically to my concern about holding on to the past, about my concentration on ancestors and the material things that remain from their lives. She wrote that maybe holding on to what remains from the past is not a problem to be solved. "Perhaps it is about 'gravitas' and carrying the past into the present and future. It's okay to have the memories. We're better off

with them than without them." My daughter added that seeing these material remains from the past has made her more connected than ever to our family and its history. Yes, let there be gravitas, with some light showing through.

Nothing is ever lost. The atoms in the creation of the universe are the atoms that are with us still, within us, and in each breath that we take. There is everlasting life in the family scrapbook.

***The winter sun is now on a higher course*** through the southern sky. Even with snow banks piled high, we know by the calendar that winter is waning. On another waning winter day, long ago in March of 1941, I was lying on the davenport in the living room of the farmhouse being carefully tended by my mother. This was my first year at Dunham School, the one-room school a mile and a half from the farm, the school that my father and his father had attended when they were my age. On the cold

## A Lifetime Burning

and dark winter day, during the afternoon recess, we students were skating on the pond in the Duesterbeck field behind the schoolhouse. Mischievously, I yelled across the ice to Bob, "Bob loves Pearl!" In a flash, Bob turned and raced toward me, and I was thrown down on the ice. Later in the week the newspaper reported the accident: "Earl fell down and another boy fell on top of him with the result that Earl had two fractures, one above the ankle and one below the knee in the right leg." My mother and father had been called, and I was taken to the clinic in town to have the bones set and a plaster cast fitted to my leg. I spent the next six weeks on the couch in the living room. I remember the great comforts given to me during those weeks.

My mother helped me keep a scrapbook, now among the artifacts recovered from the attic of the farmhouse. We had pasted the many cards received from classmates, neighbors, and relatives. The printed messages on cheerfully designed cards

## Richard Quinney

wished me well: "Get Well Soon!" "Hope You're Better in a Jiffy!" "Sorry to Hear About Your Bad Luck." Pictured on the cards are flowers, cowboys and horses, puppies, bluebirds, rabbits and squirrels, chickens and ducks, kittens, and various other wounded animals on the mend. Then there are the letters from my teacher, Miss Roeker, with lessons to be studied at home. Fellow students, just learning to print and to put words together, sent me notes of encouragement on tablet paper. Lawrence wrote, in words now fading: "You are a good boy to stay in bed for such a long time. I have a little red cow. My dog is black and white. His paws are white. His tail is white." Lawrence named him Tippy.

And, oh, the visitors that came to the house to see me during those weeks! My mother made a list, and I now count a total of 108 visits to our living room during the six weeks. Some of the visits were of neighbors or relatives that came many times, and some of the visits were of several people together. Often gifts were brought to me. The list also survives: fifty cents, candy, colors,

### A Lifetime Burning

books, oranges and apples, cookies, puzzles, pencils, gum, a picture machine from fellow students at school, clay, and fruit cake. My grandfather made a bed desk for me. I have retrieved it from a dark corner in the attic, and it now rests in a dark corner in my garage.

Never has such care been given to a boy with a broken right leg lying on a sofa for six weeks in winter on a farm in Wisconsin. Nor has such comfort ever been given to anyone under any circumstance. A loving mother and father and brother seeing to my welfare daily. A community of fellow students, neighbors, and relatives ready to rally to the needs of one of its members. That I could be so lucky, even with broken bones, at such an early time in my life.

***A dusting of new snow*** during the night. The lyrics from the song "Lately" fill the room, Lucinda Williams singing, "I only know, I miss you so, lately."

A box of letters brought back from the attic of the farmhouse. All the letters, still in their original envelopes, are dated

## Richard Quinney

1952 and 1953. This is the year I left the farm, for the first time in my life, to live away from home, from my family, neighbors, and high school friends, away for my first year in college. I had decided to attend Carroll College, forty miles northeast of the farm, in Waukesha. One reason for selecting this college was the familiarity of the road and landscape leading to Waukesha. When I was growing up, especially in winter, I would ride with my father to purchase parts for farm machinery. The shop for the Oliver tractor dealer was located at the bottom of the hill that led up to the college. There would be something familiar to see and remember during my first year away from home.

    The box contains the letters I received during my first year in college. My first month away from home my grandparents write to me letting me know they miss me, and they offer good-hearted advice: "We hope you make a name for yourself.

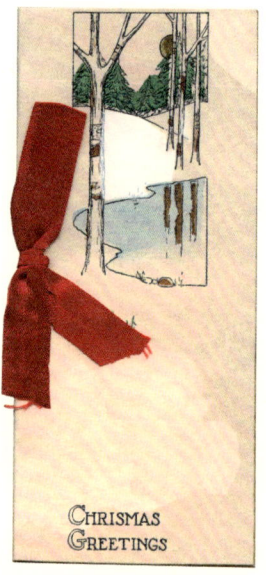

## A Lifetime Burning

Always be a good boy and stay in good company. Bad company has ruined many a young person — but we know you have too much sense to be led astray." My mother's aunt and uncle, Elsie and Lloyd, being well acquainted with adjustment to college from the experiences of their own children, write what must have been a very comforting letter. There is the information about what is happening at home and suggestions on how I might get established in a smoothly running routine, and have some time for extracurricular activities. For eight years I sat at the piano in Elsie's living room for my piano lessons, and she knows me well. "Thank you for the letters. When you spend so much of your time studying and writing, it means quite a little to write letters." Aunt Elsie tells me to come and see them whenever I am home and have the time — though she knows the time goes so quickly.

I read with great care the letters my mother wrote to me that year. The letters are signed as they would be signed for the many years to follow, the rest of my life away from home, "As ever, Mom." The letters bring news of all that is happening on the farm as the seasons pass. Each letter brings news about my father and brother, about repairs being made, the harvesting of

the crops, trips to town, visits with friends and relatives, meetings at the Grange and at Dunham School, colds and flu, Sunday mornings at church, the sale of my Duroc hogs, the birth of lambs, and reports of the trips to visit me at college.

My letters are being received and read at home. My mother writes to me, "We sure do appreciate your letters. I dash out every morning after the mailman comes to see if we've heard from you." My mother usually writes to me in the morning after she and Dad have had their breakfast, after the milking of the cows, and before she goes out to wash the milk pails and feed the chickens. Throughout the letters there is the constant encouragement and support my parents always gave me — "We have absolute faith in you and know that you will be a success and make the most of every good and fine opportunity given you."

Friends from high school write letters to me describing how difficult it is to be away from home. A friend at a university several states to the west begins his letter: "I haven't found this private life too much to my liking so far. I am thoroughly convinced that my appreciation of being on my own will increase with the time spent here. I was really homesick for the first days here. I went around with a small lump in my throat for a while. I never expected to feel that way." Mentioned several times in his letter is the fact that he is missing his high school girlfriend who is attending another college.

The box is filled with the letters I received daily from my girlfriend. She is in her senior year in high school, and next year we will be together in college. Every night we write to each other. Her letters to me, in retrospect, are a record of our life during that year. The letters express the great care we gave to each other, and the need that we had for each other. Our songs were

## A Lifetime Burning

the songs of romance: "Stardust," "You Belong to Me," "Wish You Were Here," "I'll See You in My Dreams." Good night. Our love will last. For fifty-five years the letters — the artifacts — have rested in a box in the attic. The life of that other time continues in the present moment.

**This self, this creation that I know as myself,** is not mine alone. I am, we all are, a complex of interdependently related forces and events all in a state of flux. In the Buddhist understanding of the self, we have no existence, no substantive self, separate from others. We are who we are because of our parents, our friends, and people we may never know, all who contribute to our birth and survival. We are constantly influenced by and exert influence on others and the world around us. Our interconnection — our unsubstantiality as separate beings — is social (relating to other people), physical (relating to all things material), psychological (relating to mental and psychic phenomena), and cosmic (relating to a universal consciousness). And everything is always changing and becoming something else. No wonder we look into the mirror each morning and wonder who we are, who we might be, as we awaken from another night of sleep. Each day is a day of dependent origination.

This truth is captured in the often-quoted lines from William Faulkner: "The past is not dead. In fact, it's not even past." The past is happening over and over again. It is in our present. Everything we do, unless you can imagine inventing something completely new, is connected to what has happened before. There is no present moment, in this sense, that is independent of the past. There is no substantial present that is separate from the moments of the past.

The present is what we are doing now, in this very moment. The past may be past, as Mary Oliver writes in her poem

## A Lifetime Burning

"Mornings at Blackwater." And "the present is what your life is." Your life cannot be separated from your past, and the past of all others, living and dead, animate and inanimate. Oliver ends her poem: "and put your lips to the world. /And live your life." The tension, the supposed separation between the past and the present, is purely in our thoughts.

This present I am living, this present that is fast moving to another present as I complete this sentence and begin the next, is fast becoming the past. Who I am, now, and in the "nows" of the future, is a composite of the letters I wrote and the letters I received that first year in college more than half a century ago. The letters in the box in the attic, the letters reopened and read again, show that I have never been alone. Many of the letters, in direct response to my letters, confirm the part I have played in the lives of others. We are of one mind and body.

*I have ceased prophecy*, great knowledge, the understanding of mystery, and the faith to move mountains. The treasure is simple and close to home. Paul's hymn of Christian love, written in the letter to the church at Corinth, continues as the ideal for loving others. Inspired ultimately as the love of God, this is the love we can have for one another here on earth. Beyond belief, beyond the faith of the Corinthian community, is the daily love that is the greatest beauty of all.

Behold, in the scrapbook my mother kept is the newspaper clipping reporting my baptism on a Sunday morning at the Delavan Methodist Church. I can imagine being held by my father and mother as the ceremony took place. In a cardboard box found in the drawer of the buffet in the dining room of the farmhouse, pressed and folded, is the outfit I wore for the baptism, with the note attached: "First dress Earl wore on the day he was baptized, June 17, 1 month and 1 day old." Here, I know love as the greatest beauty, years and lifetimes ago, and with us still.

When waiting for my calling, assuming that a call would someday become evident, I thought about being a minister. During childhood and adolescence I had remained close to the church and its organizations and activities. One summer while in

### A Lifetime Burning

college, I attended a religious retreat for youth in South Dakota. My spirit was not moved when, sitting around a campfire, we were asked to come forth and be saved. The spirit may not have moved within me, but I had an epiphany that shaped the rest of my life, the realization that my spiritual path would be beyond the church.

In midlife, after years of giving little thought to religious matters, I started to give more explicit attention to the spiritual life. I earnestly thought about and personally sought the meaning of life, and the meaning of my life. Insights were found in the range of traditions: neo-orthodox theology, liberation theology, Taoism, and Buddhism. The whole world—all of nature including our human existence—had become a spiritual sanctuary.

Nothing could be devoid of the sacred. The material of this world is as much of a sacrament as anything that is sanctified by any religion. The spirit within me rises and falls with the attention and reverence I give to each moment. And constant is the knowledge that once I was held by a mother and father who loved me very much.

### Richard Quinney

**In our family,** I can't remember ever hearing anyone verbally express the phrase "I love you." There didn't seem to be the need to state our love in words. The frontier psychology of not showing emotions still held firm out in the country. Our sense of love, in terms of I Corinthians, was not thought of as an emotion. Verbal expressions of emotional love, as a feeling, were confined to the romancing of our sweethearts. To this day, my verbal expressions of love are carefully reserved.

In my mother's scrapbook, I find cards she gave to my father and cards my father gave to her that are signed with words of endearment, of loving the other more than all else in this world, and a card signed "All I want is your love and you." Following this last card are two clippings from the local newspaper, both containing interviews with my father. I had no idea that my father had ever been interviewed, and I am grateful to my mother for saving them, for the world to see after the passing of many years. The first interview took place one morning in July of 1936 at the dairy plant where my father had just delivered the cans of milk from his daily route. The second interview was conducted in the fall of the same year in a field at the farm.

### A Lifetime Burning

The reporter was gaining first-hand accounts from farmers whose crops were suffering from summer's heat and the lack of rain. Oat crops were not ripening under the searing rays of the sun, and grain fields were being cut or turned to pasture. Floyd Quinney had these words to say, when pressed: "My cows are slumping off fast, and I have been forced to feed silage ... small grain outlook is extremely dubious ... had a good crop of hay ... haven't been able to do much with my horses ... wish I had a tractor ... it will take two or three days of good soaking rain to restore vegetation." I treasure my father's words, hard times or good times.

For the second interview, the reporter, accompanied by another newspaperman, had driven out to the farm from Delavan. They walked into the field, where my father was plowing with the team of four horses, and they talked. Back in his office, the reporter wrote the article, giving it the title "Farmers You Know." My mother clipped the article from the newspaper and pasted it in the scrapbook, for anyone to read who might pass this way again.

## Richard Quinney

*If you live anywhere near the town of Sugar Creek, and have not heard of Floyd then you have been passing up a most likable fellow with whom you ought to make an acquaintance.*

*Floyd lives about five miles from Delavan. His farm is only moderate in size, consisting of about 90 acres, of which 60 are on one side of the highway, and 35 acres on the other. A winding road, which makes a pleasant drive, leads you to the farm, and Floyd's only kick against it is that the snow always lodges there in big heaps during the winter.*

When your touring Enterprise field men arrived in Mr. Quinney's front yard one day recently, they spotted Floyd out in a nearby field finishing up some fall plowing. Plowing on the Quinney farm is all done by horses, and on this particular day, four of them were hitched abreast to a riding gang plow.

Floyd has been a loyal booster of the Enterprise since last summer, when he broke into print on a hot July afternoon, as he expressed his opinion on the drouth to a reporter. Still he was reluctant to be interviewed. "I don't believe I've done anything so wonderful. Why don't you write up some of the bigger farmers around here?" These were of his statements before he broke down and answered a few questions about his neat and well-kept homestead.

Eighteen cattle, 14 of them Jerseys, and four of them Holsteins are found on the Quinney farm. The Jerseys are of high grade, and their milk has a butterfat content of about 5.5, which is almost comparable to Phil Steffanus' Fawn Dawn herd. The herd sire appears to be a vicious looking animal, and when strangers come within hearing distance, he becomes excited and kicks up quite a fuss. This happened during the visit of the Enterprise men, but a few reassuring words from Floyd sort of calmed him down.

## A Lifetime Burning

*Floyd works hard. He does most of his work alone, and manages to get ahead somewhat every year. His harvests for this year were satisfactory to him. All of his corn crop went into his silo, but he's not so disappointed over that. His grain crop was good, and he's well stocked on feed for the coming winter. Buildings and machinery reflect evidences of careful attention. The big barn is decorated with a coat of red paint, and all of his machinery is carefully stored away out of the weather when not in use. Floyd believes "it's the abuse rather than use" which is destructive to machinery.*

*The dwelling on the Quinney farm is a neat homey bungalow type of structure, painted cream color and trimmed in green. There is a neat lawn and farmyard, which adds to the attractiveness of the place.*

*Work is made easier by the use of electricity. "We hooked onto the high line a short time ago," said Floyd. "Formerly we had our own plant, but we like this arrangement much better. It is capable of doing lots more work."*

## Richard Quinney

*Right now Floyd is worrying whether or not the highway to Delavan will be clogged with snow during the coming winter. Last year he had to use a round-about trail through the fields in order to get to town. Open highways are mighty important, because he hauls milk every day, and hopes that he can use his new truck instead of hitching his team to the bobsled.*

*Some of the neighbors of Mr. Quinney out in Sugar Creek are Morris Johnson, Oscar Kittleson, and Roy Williams.*

*The future of the Quinney family seems to be well assured, since Floyd has two young sons, named Earl and Ralph, who comprise the new generation.*

**The two young sons of Floyd and Alice** could not assure the future of the family farm by staying on the farm, as the reporter suggested they might. My father and mother did not

## A Lifetime Burning

encourage us to be farmers. Although we left the farm to pursue other lives, we have sought to preserve the farm by keeping it in sustainable production, hoping for a future. One could say, out of love.

***T**he sun has crossed directly over the equator* on its apparent trip northward. Equinox — equal nights and equal days. A promise of the change in seasons as the Earth tilts on its axis. There is less of a slant as the rays of sun pass through the sky.

I do find more references to love as I sift through the artifacts from times past. In the scrapbook, my mother pasted a clipping on love being central to the home:

> *So what does it take to make a home, this thing that we all crave more than anything else in life and which we all try to fashion with our bungling hands, and which sometimes turns out a success and*

## Richard Quinney

*sometimes a failure? It takes, first, love. The love of a husband and wife, the love of parents and children, the love of brothers and sisters. Love is the first on the hearthstone, the light in the window that makes a home warm and bright. Without it everything is cold and bleak and as inhospitable as the tomb. There is nothing at which you can warm yourself. The married couple that has only a lukewarm liking for each other and whose children are unwanted cannot make a home. They only build a house.*

Such is the importance of the printed word when speaking the words might be uncomfortable. On the same page, saved in cellophane, is a pressed yellow flower — "My birthday gift from Floyd, April 29th, 1936."

Trivial, then, is the report in the morning paper that predicts the Earth's demise in 7.59 billion years. Planet Earth will be dragged from its orbit to a rapid vaporous death. Some things, we may believe — of the heart, the mind, the spirit — are unbounded by earth and sky, and will survive. My material home on the farm has already been dragged from its orbit. But much remains.

## A Lifetime Burning

***S****till this searching,* this attempt to put into words, this vain hope that I have of finding peace through understanding. In the writing, I have the sense that something may be resolved, however momentarily. At least I know that I am alive when I can write. Maybe I am the most aware in these meditative moments of writing. I like Robert Frost's comment on writing poetry: "A poem begins with a lump in the throat; a homesickness or a lovesickness. It is a reaching out toward expression, an effort to find fulfillment. A complete poem is one where an emotion has found its thought and the thought has found the words."

I sit at the table in the morning with a lump in my throat—and some relief is found in a contemplative thought followed by a few words. Homesickness is assuaged—once again—when home is found, remembered, recognized. Frost added a comment on life itself: "In three words I can sum up everything I've learned about life: it goes on."

The cardboard box filled with the contents from the kitchen drawer at the farmhouse contains the recipes my mother used throughout her life on the farm. Some of the recipes—on cards, scraps of paper, and in notebooks—go back to earlier generations. The week my mother died, at the age of ninety-two, she baked cookies to be taken to the nursing home in Delavan. The ceramic and wood rolling pin that she used when baking cookies I have sent to my daughter for safekeeping.

Chocolate drop cookies baked by my mother, with the hickory nuts that we cracked in the basement on winter nights, were always a favorite of mine. The dough was dropped onto the pan and baked in a medium oven. Her many recipes include the devil's food cakes that she baked throughout my childhood. From the drawer in the kitchen, I have taken one of the recipes.

## Richard Quinney

Tonight I will bake a cake. From years of experience, my mother knew the size of baking pan, the temperature of the oven, and the baking time. I will have to take chances in my first attempt at baking a devil's food cake.

Someday I will try other recipes. I will think of my mother standing at the kitchen table rolling out the dough for another batch of cookies.

**The letters that I have taken from the farmhouse**—from the dresser drawers and from the boxes in the attic—could fill notebooks for the rest of my life. I wrote home regularly to my mother and father all of my years in college. While away at graduate school I sent letters weekly describing my studies and my first attempts at being a teacher. Not far below the surface of the writing is a continuing sadness about being away from my family and the farm. And then a letter home announces that I will be married with the coming of summer. My mother's letter to me, in response, expresses surprise and concern.

## A Lifetime Burning

The letters over the next forty years—to my parents and from them to me—traveled thousands of miles. Grandchildren were born, graduations and marriages were celebrated, vacations were taken, gifts were given and acknowledged, relatives and friends were visited, changes came to the farm, neighbors and relatives passed away, I moved from one place to another, and my mother remained on the farm for thirty years after my father suddenly passed away on a November day in 1969.

Among the letters in the farmhouse were those my daughters wrote to Grandpa and Grandma and later to Grandma alone. I have returned the letters to my daughters so they will know the care they gave to their grandparents. Letters would stop coming and going only with the death of my mother after seventy years of living in the farmhouse. The lives so well lived survive as recorded in the fading ink on the aging paper of the letters. The letters remind me, rekindle in me, whenever needed, the knowledge that a lifetime burns in every moment.

**My exploration of the past** in the writing of memoir alerts me to the facts of impermanence and my connection to all others. Hopefully, my continuing explorations are leading me to a better and more abiding grasp of impermanence and my connection to everything else. A better understanding of how this particular human life fits into the larger scheme of things.

My teacher from afar for many years, Thich Nhat Hanh, keeps reminding me to stay focused on the present moment. Not to dwell in the past, not to remain in sorrow and regret, but to be in touch with what is happening now. He writes clearly, in *Touching the Earth*, about the dangers of dwelling in the past, of regretting what we have lost:

> *This robs us of the opportunity to be in touch with the refreshing, beautiful, and wonderful things that could nourish and transform us in the present moment. We are not able to be in touch with the blue sky, the white clouds, the green willow, the yellow flowers, the sound of the wind in the pine trees, the sound of the running brook, the sound of the singing birds, and the sound of the laughing children in the early morning sunlight. We are also not able to be in touch with the wonderful things in our own selves.*

I hope to dwell in the present, even as I explore how my life in the past is shaping my life. I know that to remember the past is not to live in the past. Remembering is a vital part of living in the present. Time present is the only time we really have.

Being aware of impermanence, interdependence, and the reality of the present moment can lead us to the truth of no birth, no death. Arriving, as Thich Nhat Hanh instructs us, "at the deep realization that all that exists has the nature of no birth and no death, no coming and no going, no being and non-being, no permanence and no annihilation." A good place to be as we live our lives each day.

## A Lifetime Burning

**In my mind,** my father stands at the front of the sleigh, reins in hand, the breath of the horse steaming from nostrils into the winter air. My brother and I sit on the milk cans, ready for the ride in the deep snow to the main road. We will be on time for the first class of the day at Dunham School. My father, as usual on winter days, is wearing his brown canvas coat, trimmed in leather and lined in sheepskin, with the large wool collar. The winter coat would hang in the attic long after my father was gone. Each time we climbed to the attic we would think of him, reins in hand, about to take us through the snow on a winter morning.

As we cleared the attic, I suggested that we try to preserve the winter coat. We took it to the cleaners, and when we picked it up we found that the coat had not survived the cleaning. We were handed the shredded coat wrapped in a plastic bag. But I will keep the shredded coat for a while — as a lesson that there is little in this material world that can be physically preserved. The things of the past exist as long as they are remembered in our hearts and minds.

**W**hen all is said and done, after all the gaining and the spending, after all this trying to make sense of life, I know that I must ultimately surrender. Everything that I cherish today I will someday be separated from. Only what I have given to others will last and not be lost.

I will take courage in the later life of one of my Holloway ancestors in Exeter, Canada. Well into her nineties, Reta and her husband, Jim, had to move out of their house and into a retirement center, giving up most of their possessions. Later, at the funeral for Reta, her niece delivered a eulogy that recognized the loss of their material possessions: "She taught me the meaning of letting go. It must have been painful for them to part with so many of their valued lifelong possessions, but the ease with which they pared down this former life to make way for their new life amazed me." A new life, life itself, was the treasure.

### A Lifetime Burning

**The cover of the childhood scrapbook** that my mother gave to me shortly after I was born is embossed with the moon looking down at five owls perched on the branches of a tree. Bats are flying in the night. There is a four-line verse in the lower left corner of the scrapbook:

> Five little Owls sitting on a tree,
>> Up rose the fair Moon, wonderful to see.
> "Fly away! Dear Owls, fly away to me."
>> "Thank you very kindly we'll stick where we be."

Where we belong, here on earth. This is the only home that we have, and the home where we will always be.

## Memoirs by Richard Quinney

*Journey to a Far Place*
*For the Time Being*
*Borderland: A Midwest Journal*
*Once Again the Wonder*
*Where Yet the Sweet Birds Sing*
*Of Time and Place: A Farm in Wisconsin*
*Tales from the Middle Border*
*Things Once Seen*
*Field Notes*

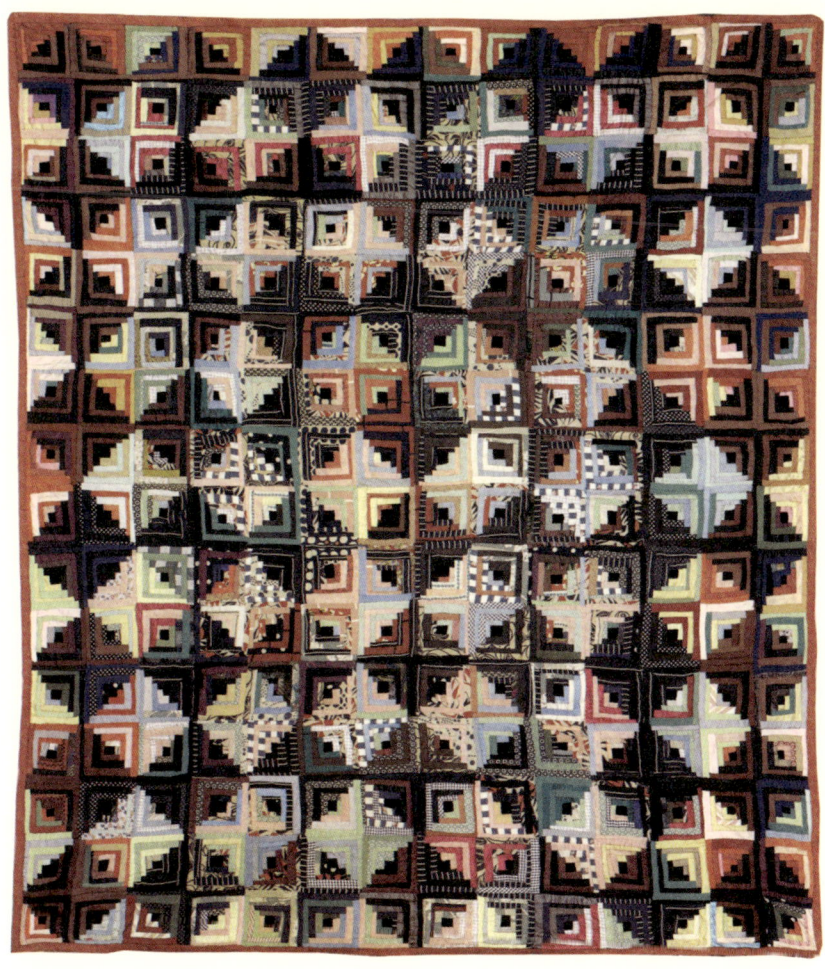

*Quilt made for Floyd Quinney in 1929 by his aunt Kate Quinney*